THE INNS

OF

BURFORD

PART ONE:

THE HIGH STREET

Raymond Moody

1996

Text and illustrations
Copyright © Raymond Moody 1996

Third printing 1997

ISBN 1 901010 01 5

Also in this series:
Burford, the Civil War and the Levellers
The Great Burford Smallpox Outbreak 1758
The Inns of Burford Part II: Beyond the High Street
Forthcoming:
Roads and Rogues: I - Turnpikes, Traffic and Travellers 1750 - 1800
Roads and Rogues: II - Vagabonds, Villains and Highwaymen
The Burford Year: Markets, Fairs and Festivals

FOREWORD

More nonsense is talked about the history of inns and public houses than about that of any other establishments, and for a well understood commercial reason. Other businesses boast about the modernity of their properties, but what counts with inns is antiquity. Almost every country town hotel in its advertising material aspires to be a coaching inn and if, by virtue of a fragment of an oak beam or a rough stone wall, it can claim to be mediaeval (elsewhere an epithet of denigration), so much the better. The claim of age in the buildings may be well founded. Few towns are older in their structures than Burford, and before the days of the JCB and the tipper truck, not many buildings were swept away so that a fresh start might be made upon their site. It was usual to alter and adapt to a new purpose as trades came and went. But the building is not the business, and only one of Burford's inns to-day has been trading continuously since before 1700.

This booklet is based on the many hundreds of old Burford property deeds and leases that I have read over the years, on county newspapers (for much reading of which I am very grateful to my wife), on parish and town records, on charity documents, on tax returns and court records, on wills and probate inventories, the census, trade directories and other sources. The large inns are well documented, but some alehouses must have escaped me, and there are a few elusive establishments of the past that I can name but still not place on the map. They are named within, and if anyone can help me with these I shall be very grateful.

R.M.

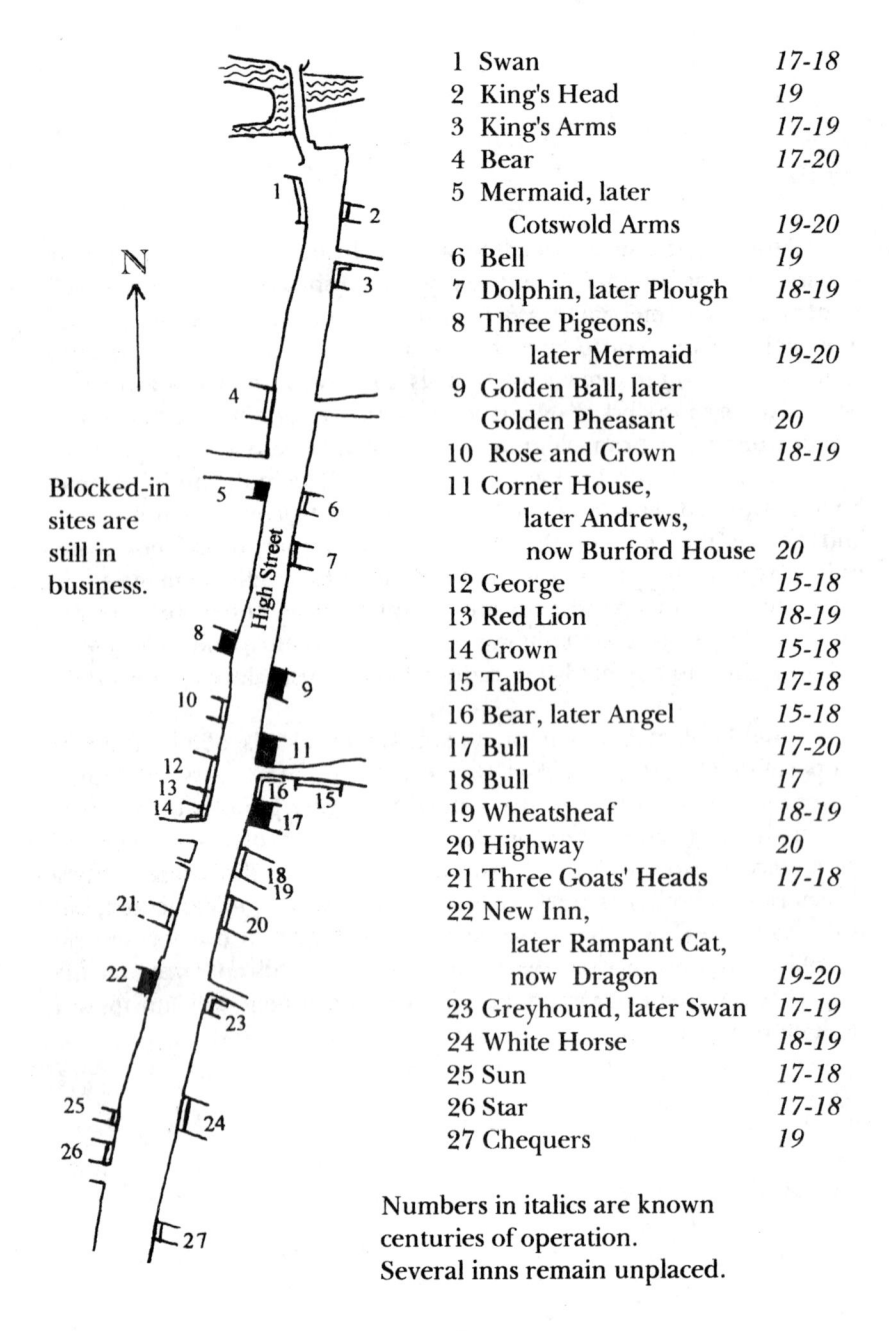

#	Name	Centuries
1	Swan	*17-18*
2	King's Head	*19*
3	King's Arms	*17-19*
4	Bear	*17-20*
5	Mermaid, later Cotswold Arms	*19-20*
6	Bell	*19*
7	Dolphin, later Plough	*18-19*
8	Three Pigeons, later Mermaid	*19-20*
9	Golden Ball, later Golden Pheasant	*20*
10	Rose and Crown	*18-19*
11	Corner House, later Andrews, now Burford House	*20*
12	George	*15-18*
13	Red Lion	*18-19*
14	Crown	*15-18*
15	Talbot	*17-18*
16	Bear, later Angel	*15-18*
17	Bull	*17-20*
18	Bull	*17*
19	Wheatsheaf	*18-19*
20	Highway	*20*
21	Three Goats' Heads	*17-18*
22	New Inn, later Rampant Cat, now Dragon	*19-20*
23	Greyhound, later Swan	*17-19*
24	White Horse	*18-19*
25	Sun	*17-18*
26	Star	*17-18*
27	Chequers	*19*

Blocked-in sites are still in business.

Numbers in italics are known centuries of operation. Several inns remain unplaced.

"Shall I not take mine ease in mine inn?" (Falstaff: Henry IV, Pt.I)

SOME MEN look for a convivial drink beside a fire of logs in a wide stone fireplace; some come for a fine meal in a warm room; some want a comfortable bed for a night before they go about the next day's business; some stay many days for rest and release while some stop only for a brief lunch before they take up their toil again; and there are those who come to a house that was already old when their great- great-grand-fathers were young men, to imbibe the flavour of old England, mellow stone, oak beams and good cheer. All these come to an inn and do not go away disappointed, for inns have a sure and ancient place in our life, and nowhere more than in our country towns. Inns are far older than village halls: the officers of the parish or town met in them, merchants sold goods in them, bargains were struck in them, auctions took place in them, "particulars could be obtained" from them, plays were performed in them, in the days of growing fashion balls and assemblies were held in them, doctors practised at them and even monarchs slept in them from time to time. Stage coaches and stage wagons changed horses, picked up and set down passengers and delivered goods at them, cock fights and prize fights were held there and hunts met outside. Well-known stallions were at stud at inns. Horses, chaises or coaches could be hired from them. The inn divided the public life of the town with the church. One catered for the needs of the soul; few needs of the body could not be met at the other. The keeper of a great inn was a leader in the commercial community and the varied functions of his establishment gave him power in many directions. Wealthy merchants might aspire to be innkeepers and innkeepers often became wealthy merchants and served as bankers in their town. Burford was chartered as a market town nine centuries ago and inns have always been a great part of its life.

Hotels are late in the field, for the name was adopted in the 1770s by certain inns needing a new image to retain their customers as some of their many functions were drawn off by the rise of specialist entertainment or lodging or eating establishments. Suddenly hotels were the vogue: - "*in 1763 there was no such place as an Hotel: the word indeed was not known or only intelligible to French scholars"*. In 1817 it was thus defined: *"Hostel, hotel - a genteel inn; this word is now universally pronounced and written without the s."* As the functions of an inn contracted so it became an hotel, a place primarily concerned with accommodation and meals, much as a public house has developed from an alehouse.

The great town inn was the summit of a pyramid of which the alehouses were the base. Alehouses for long demanded no specialised premises. All that was needed was a room and a bench and a fire in winter. An artisan seeking more money, a widow or a wife in search of an income, would open their premises as an alehouse. It could disappear as casually as it came, leaving as little trace of its passing as its humble customers. Beer made with hops was not introduced into England until Tudor times and then its favour spread slowly. The earlier liquor of the alehouse was easily brewed in a kitchen. The alehouse had at first no name or sign as we know it, but hung out an "ale stake" with a bush on it when the brew was ready. Perhaps "good ale needed no bush" because the word went swiftly round; and it would need to, for the brew would not keep. Estimates suggest that across the country there was an alehouse to every ninety persons and in a market town like Burford there would have been still more. At any time there may have been twenty-five such establishments here and there is only a passing reference now and again in the records to witness to an institution that began far back in the Middle Ages and merged into the public house of today almost in living memory.

Parts of the Manor Court rolls of Burford have survived from the middle of the sixteenth century, and we have the names of those who presented themselves as brewers, as keepers of common hostels, as alehouse keepers, and as vintners at the very end of the reign of Henry VIII. The vintners, Richard Hannys and Agnes Green in 1547 and John Hannys and Richard Charley (who is also named as an innholder) in 1549 we can identify as the keepers of the two most important inns, the Bull and the George. The list is likely to be incomplete for only the more respectable of premises will appear but even so we have in 1547 twelve alehouse keepers and in 1549 nine brewers. All those who can be identified from their wills have other occupations, and are not known primarily as alehouse keepers or brewers.

Almost a hundred years later an incident in a Burford alehouse surfaced in the Archdeacon's Court. One day in late November 1632 Laurence Griffiths the curate of Holwell, some three miles away, was in Burford on business and became delayed. He went to the house of Robert Aston "*being an inn or alehouse*" asking, so he said later, for lodging. This fact, together with the alternative description as an inn, marks out Robert Aston's establishment as being of the better sort. He found good fellowship well advanced inside the hall and, perhaps more willingly than he later admitted, joined the party, most of whom were unknown to him. Trouble began when he found himself without the means to pay his share. One Simon Partridge of Burford pulled off his coat, one Drewett pulled off his boots. Then, he complained, he was by Drewett "*in violent, contemptuous and scornful manner putt into a tubbe or vessel and by him bid to preach and after the said Drewett ... put a linnen cloth over his shoulders*" bidding him do penance, and then his face was blacked with charcoal before they allowed him out of the tub. Three and a half centuries later one wonders if the parson did not protest too much his innocence in the whole affair. More interesting is the fact that the Robert Aston of the inn must be the Robert Aston who was buried in 1655 and was the ancestor of the Aston and Tash dynasty of innholders lasting for a century and a half in Burford and which ran both the George and the Bull in turn. They were a prolific family: this Robert had ten children, and his son Robert also ten, all of whom outlived him.

The Crown.

Mediaeval documents are scarce and not until 1423 do we read of a Burford inn, though they surely existed. There is a deed of that year conveying a property called the "*novum hospitium angulare*" - the new inn on the corner. It must have been even more obviously on the corner before the Tolsey was built in the roadway. It was owned by Thomas and Christiana Spicer, members of the family whose memorial brasses are in Burford Church. It was to pass successively to their three children and to their grandchildren if any and, failing them, to the Parish Church. In 1464 it became church property, the gift of the heir of the Spicers' second son. By 1507 it had acquired what we would recognise as an inn name and was called the Crown.

The former
Crown Inn,
now
the Pharmacy.

There is an interesting parallel here with the building on the corner of Ship Street and the Cornmarket in Oxford, a Jesus College property now mostly the Laura Ashley shop. John Gibbes, vintner and five times mayor of Oxford, built his "*novum hospitium*" there in the 1380s, and this too became known as the Crown by 1430 - "*novum hospitium ubi corona est signum*" - when it was owned by William Brampton, the Burford man who was Mayor of Oxford seven times between 1420 and 1439. We shall meet the Bramptons again at the George.

The Crown was originally a large property extending along Sheep Street, with a close adjoining and an internal courtyard. By 1544 a house had been built behind the Crown on Sheep Street, and before 1590 the Crown had been divided into three parts, about the boundaries of which there had been disputes, for in 1598 detailed descriptions of the parts were drawn up. It is difficult to relate these to the buildings of today. One third - apparently the front - had been owned by Richard Hodges of the George, and then by his widowed daughter Alice Reynolds (almost certainly the widow of the Richard

Rainoldes whose tomb is in the south aisle of the church) and was used as a shop. The other two parts remained with the trustees, but at some point the parts were united again.

From 1625 we have the churchwardens' accounts. In that and the following year they "*receaved of Steven Scott for one yeares rent for his nowe dwelling house called the Crowne in Burford xxvj sh.*" Stephen Scott's daughter Avis was married to Robert Collier, vintner, of the George, tying him into one of Burford's innkeeping dynasties. Scott died in 1627 and attached to his will is a probate inventory making it plain that the Crown was not merely his dwelling house but an inn. The contents of the hall, the chamber called the Lion over the kitchen, the chamber over the hall, the green chamber, the great chamber next the street, the Rose chamber, the little chamber, the chamber over the entry, the cellar, the taphouse, the kitchen and the brewing kitchen are all listed in detail. The total valuation was £176. 12. 6d, a tidy fortune then. His widow Susan continued to pay rent on the Crown until 1630, after which the accounts only give the total of rents, though in that year the property was leased again to John Woodward, gent. at £6 per annum.

The next known occupant is John Syndrey, mercer, in 1659 and from the turn of the century John Castle and then William Castle are there, at £14 p.a. Though described as chandlers, the Castle family had medical connections, and in 1734 after William Castle's decease Nicholas Willett, apothecary, (whose family we shall meet again) took a lease of the premises. The mention of another occupation for the holder does not mean that the Crown had ceased to be an inn, but sometime in the middle of the century the inn trade ceased, and a pharmacy it has remained ever since.

The Crown had its moment of notoriety in 1649 when, after the Civil War, a mutiny broke out in the Parliamentary Army. Nine hundred troopers on their way from Salisbury to Banbury halted in Burford for the night; every available space in Burford and the surrounding hamlets must have been occupied. After covering fifty miles in the day Cromwell's men stormed into Burford about midnight and the mutiny was over. Only at the Crown was there a point of resistance, and the church register records: "*A soldier slaine at ye Crowne buried y 15 of May*".

The Bear, later the Angel.

The Crown is one of the inns we have lost. Another is the Bear, not to be confused with the seventeenth century Bear which survives as Bear Court in the lower High Street. Its twin gables, now oddly contained in a single truncated roof, faced the High Street on the corner of Witney Street next to the Bull. It is first mentioned in 1489 when it "*lately*" belonged to John Pynnock, head of another wealthy family which, like the Spicers, dominated Burford in the fifteenth century. Politics and inn names went hand in hand. The Bear perhaps took its sign from the arms of the great Earl of Warwick, the Kingmaker of the Wars of the Roses, who counted Burford among his lands from 1449 to 1471. When he fell and the Tudors were firmly established on the English throne, it was wise to change the name of the Bear to the Angel. It had become the Angel before 1539 when William Pynnock owned it and the Angel it remained, kept by Richard Hodges in 1599 and Thomas Parsons in 1652 and 1685, though we know little else about it. In the eighteenth century it was held by members of two important related Burford innkeeping families, Edward Castle and the Chavasses. Some coaches were still stopping there in 1773 but after that it vanishes, absorbed into the prospering Bull next door and the trade directories that begin at the end of the century do not mention it. After two centuries, the Angel has been re-incarnated nearby in Witney Street.

The George.

Without a doubt the Bull is the oldest of the surviving inns of Burford, but for two centuries it had a rival in the George, first mentioned in 1485, the year of Bosworth Field, but probably much older. The name is that of St. George, brought back by the Crusaders from the Middle East to become the patron of English arms. When the Hanoverian Kings arrived in 1715 and the Middle Ages were forgotten, the popular mind confused the saint with the kings, as the Mummers' plays often do, and the sign may have ended by showing the puffy features of the kings (as the Old George at Leafield) and not the saintly hero spearing the dragon. The three gables of the George still look out on the High Street and you can still walk through its yard, but now it is divided into shops and offices and the buildings of its yard have become nine cottages. William Brampton owned property on this site on the High Street from at least 1404. He was described as mercer (merchant) of Oxford in 1419 where, as we have seen, he owned the Crown Inn on the corner of Ship Street and the Cornmarket. He died in 1443. The George was still in the family in 1485 when the Tudors came in.

In the long family struggle between York and Lancaster known as the Wars of the Roses most Burford inhabitants were too busy minding their own small businesses to take a part, but when Henry VII, the victorious Tudor monarch, drew up the list of those whose property he would confiscate, he began with John, Duke of Norfolk and ended with William Brampton of Burford. The king, wishing to unite his realm, limited the list to those whose opposition had been serious. We may wonder what part Brampton played at the Battle of Bosworth. Among his confiscated property was the George and, although his sister managed seven years later to have the attainder annulled, the George had been granted to a king's servant, and it was granted again to William Gower, a Groom of the Chamber, to be held "*at the King's pleasure*". An inn like the George in a town like Burford, then among the county's four most prosperous towns, was a valuable property. It came back to local men, and to some of the wealthiest and most respected names in Burford, though in the scanty records it is not always easy to distinguish between holders under the Crown, owners in fee simple, the lessees of the property and the actual innkeepers. Richard Hodges held it in the Royal Manorial Survey of 1552 when it was the only inn named in Burford. After him came Richard Charleye, described as Gentleman, of a landed family in Lancashire whose brothers were London merchants and whose wife Agnes was sister to Simon Wisdom. It passed to the Collier family. John Collier was an important man in Burford: burgess and bailiff of the town's corporation, and churchwarden as well. Under John Collier the George was expanded. The uphill third on the High Street (the structure now with Jackson-Stops' fascia on the front) was added in 1608, standing on one of the properties known as Poole's Lands

The George Inn today.

belonging to the charities of the town. The measurements of the new building are given in the lease: 22 feet along the High Street, and 32 feet from front to back. The George already had the characteristic feature of an important town inn - an archway leading to a yard, though no Burford inn now retains the well-known detail of a gallery looking down on the yard. The yard, of course, was for the stabling, when horses were as much an essential of travel as cars are now. Horses were important to Burford in another way, for Burford Races, held on the open downs at Upton until they were enclosed, and later near Aldsworth, were famous and were to become second only to Newmarket for two hundred years. They brought rowdy society to the town. In 1626 the Parish Register has the burial of "*William Baxter, gent, killed at the George next day after the race.*"

We are fortunate into having an insight into the George that gives us a lively picture of its interior. With the wills of two of its seventeenth century holders - John Collier in 1634, and Richard Veysey in 1667 - are preserved room by room inventories, listing the furnishing, down to the iron bars in the kitchen chimney. The more important sleeping chambers are named, probably from details of the elaborate plaster decorations or paintings long since gone: the Lyon, the ffortune, the Crosse Keyes, the Starre, the George Chamber, the Rose Chamber, the Greene Chamber, the Newe Chamber, the Tower Chamber. A glance at the building shows clearly where the Tower Chamber was. There is a parlour, a hall, a taphouse, a bar or beer-room, a kitchen, two cellars, a spence, a gallery, two stables, a brew-house and an ostlery. The inventories are too long to reproduce in full, but let this room in 1634 serve as an example of inn comfort:

>   *In ye Greene Chamber*
> *Imprimis one greene Rugge & two blancketts*
> *Item one featherbed two bolsters & two pillowes*
> *Item one other featherbed one bolster and a pillowe*
> *Item one coverled and a blanckett*
> *Item one standinge bedsteed a truckle bed*
> *   five curtaines Valence and Rodds*
> *Item one tableboard sixe stooles three chayres and*
> *   Round table and a Court Cupbord*
> *Item four turkie worke cushions & three Irish stiched*
> *   a carpett and a Cupboard cloth*
> *Item an Iron backe in the Chymney*
> *   a payre of doggs and tongues*   (i.e. fire dogs and tongs)

The total value of the furnishings in this room was £17 12s 10d. John Collier's own room in 1634 gives us a view of the man who ran the inn:

>   *In his lodging Chamber*
> *Imprimis one featherbed bolsters two pillowes*
> *Item one bedstead*
> *Item three coverleds foure blanketts & two Coffers*
> *Item a presse*
> *Item a Coffer a truncke two chests a deske*
> *Item his wearinge apparell*
> *Item a gold ringe*
> *Item in money*

Total value £16. 5s. 4d. No fire place and fittings but rather more blankets and, of course, the chests and desk of a man of affairs. These furnishings must be set against the emptiness of the houses at this time, when the inventories of the ordinary merchant would list a couple of stools and count the spoons. The total probate valuation of John Collier in 1634 was £366 13s 3d and of Richard Veysey in 1667 £409 6s 6d. These figures are many times the average wealth of traders in the town. There was a fortune to be made in keeping a successful inn, but perhaps only a man already prosperous had the capital to undertake it.

Collier's daughter Marjorie married Marchmont Needham and her son, also Marchmont Needham, was the notorious pamphleteer of the Civil War, always attacking and often changing sides. Her second husband was the Vicar of the parish, the Revd. Christopher Glyn. The Civil War brought a royal visitor to Burford. In 1644 Charles I fleeing before the Parliament Army from his refuge in Oxford passed through Burford on his way toward Worcester and briefly refreshed himself at the Priory. A few days later and still retreating he was back again and this time one contemporary source records that he was lodged at the George. With his cause almost lost he must have been glad to rest his head even at a public inn.

One of the Colliers, Thomas, went off to London to make his fortune. He did not, like Richard Whittington from Gloucestershire, become thrice Lord Mayor but he did prosper and in his will of 1664 he left fifty-two shillings a year from the rents of his brewery in Shoe Lane, Holborn, to provide a penny loaf for each of twelve poor boys in Burford each Sunday after morning prayer at the church. He did the same for the parish of St. Andrew's, Holborn. In 1666 London burnt and his brewery with it. The tenant disowned his responsibilities and no more money was paid until a court sitting at the Guildhall in 1674 affirmed the charge on the new building on the site. The payment was commuted for a capital sum a hundred years ago and invested now produces more than £2.60 each year, but today it provides bread only once a year, just before Christmas, when specially baked rolls are given to the children of the Primary School and Thomas Collier is remembered together with all Burford's earlier benefactors. It is necessary to remind to-day's children that a weekly loaf could once mean so much to a poor boy in Burford.

After the Colliers, Richard Veysey took the George and kept it himself until 1667 and his widow after him. He was a member of a landed family in the area who were the owners of the George from whom the Colliers rented. Richard Veysey also issued tokens. In the middle of the seventeenth century the Mint did not much bother itself with small change, generally limiting the Royal issues to gold and silver. For the transactions of everyday life something smaller was needed and tokens, which effectively were small coins backed by the financial resources of local business, were widely used. The keeper of a good inn had both the wealth and the facility to play the role of banker. Veysey's farthing tokens had on one side *At the George* surrounding a George and Dragon, and on the other side *In Burford* around the initials R.A.V.

The Jordan family acquired the George by 1652 and after Veysey it was kept for them by Robert Aston, one of Burford's great innkeepers. This was perhaps the finest period of the George. In the lusty days of the Restoration Burford Races flourished and Charles II and his court were often in Burford for the sport. When Parliament met in Oxford, driven out of London by the Plague, the King, easily bored by the business of government, would take his pleasure in Burford. I doubt whether this King ever stayed at the George - his entertainment at a great local house is far more likely - though his presence in the town is noted in 1663 and 1681. Some of his court surely stayed here and there is a persistent folk tradition linking the George with Nell Gwynne and the King. One story runs that when "pretty, witty Nell" called her (and the King's) infant son "poor bastard" in the King's presence, the King remonstrating "Call him not so, Nell", received the reply "I have no other name by which to call him". The child was created Earl of Burford and Baron Heddington on the spot. He retained these titles when eight years later he became the Duke of St. Albans and to this day Earl of Burford is the courtesy title of the Duke's eldest son. Certainly there is truth to character in the story and these titles were indeed conferred in 1676 on Charles' six year old son by Nell Gwynne. Her house in Windsor was called Burford House and its name survives today. Six of Charles' illegitimate offspring were created Dukes.

An oddity still to be seen at the George is the verse scratched, probably with a diamond, on one of the windows facing the High Street.

Teach me to hate the Author of my wrongs,
For as yet I know not what it is to hate;
My soul engrossed by softer passions
Has not room to entertain so rough a thought.
                         1666 Samuel Pepys

Someone has added, mysteriously, in a later script:
Then you ought to be damn'd.

We have Pepys' diary for 1666, the year of the Great Fire, and he was not in Burford in that year. There is nothing in the verses to suggest a genuine connection with Pepys, and the style of the lettering does not match the period attribution. It is the fantasy of a later hand embroidering on vague tradition.

Robert Aston kept the George for the Jordans from 1674 until 1697 when he took a lease of the Bull on his own account. The next known keepers of the George have left no mark beyond their names: William Gossen and Thomas Kennett. Then came the Clare family, last holders of the George. Thomas Clare was there by 1737 sufficiently well established to be appointed among the new Charity Trustees. In 1751 he was a Churchwarden, and he was there through the terrible year of 1758 when one-eighth of the inhabitants were killed by the small-pox, and he was a signatory of the letter in September declaring that the epidemic was over. He died in 1772 and his son, another Thomas, in 1779. His two daughters, Sarah and Winifred Clare, then took over the running of the inn. They had some dangerous customers, for the notorious highway robbers, the Dunsdon brothers, Tom, Dick and Harry, were drinkers at the George. In August 1784 two of them came there again, but not to enter. They waited silently outside in the cart, their feet over the tailboard, while the carter who had brought their bodies from execution in Gloucester jail to be hung in chains on Shipton Down went in to wash the road dust from his throat. More respectable was Colonel John Byng who came to the George in July 1785. He arrived on fair day and found the place crowded, but coming back a day later found it "*a calm scene of civility*" though he only pronounced the inn as "*tolerable*".

The end came in 1800. I have a mental image of a respectable faded old-fashioned establishment kept by the two sisters, then 66 and 69, while on the other side of the High Street the Bull with the go-ahead Mr. Stevens in charge was smart, fashionable and popular. That may not be entirely fair, but in 1800 the Oxford Journal carried the announcement that Sarah and Winifred Clare were giving up the George in favour of Mr. Stevens at the Bull. A sale of all the furnishings and fittings of the inn followed:

Capital Sale of Household Furniture, etc. etc. on 30th July & three following Days ... All the household furniture, four hundred ounces of modern Plate, and a Variety of Plated Goods, sixty Pair of very excellent Sheets, sixty very beautiful Damask Table Cloths of large Dimensions with Napkins to correspond, and other Bed & Table Linen;a great Variety of foreign China & Glass; an excellent Mangle, nearly new; ... the Property of Messers. Clares ... who are retiring from the Public Business in Favour of Mr. John Stevens, at the Bull Inn. The Furniture, which is in excellent Preservation, consists of upwards of thirty Bedsteads, with Moreen, Cotton, Dimity, Check and other Furnitures & Window-Curtains to correspond; thirty prime seasoned Goose & other Feather Beds and Matrasses, with most excellent Bedding; Wilton & Scotch carpets; Pier and Dressing Glasses; a Great Variety of Cabinet Work in Mahogany Chests of Drawers, Night Tables & Bason Stands; Dining, Pembroke, Tea, Card & Dressing Tables ... about four Hundred Weight of very excellent Pewter ... to commence each morning at half-past ten.

I wonder where all that good eighteenth century furniture is now. Was it discarded by Victorians who found it dull and old fashioned, or has it returned unremarked in recent years through Burford's many antique shops? Burford was not in decline in 1800, but the greatest days of coaching were still in the future, and from the wording I can only conclude that Mr. Stevens had bought out the Clare sisters. There may not have been trade for two large inns both pursuing the same type of business. For a few months Mr. Stevens ran the George as an annexe to the Bull - there were functions in the "Union Room" - but it soon closed permanently. When next we hear of the George John East is operating a stone mason's business in the yard, ancestor of a business that lasted in Burford until the 1970s. To-day the road front of the old inn is divided into offices and shops, while in the yard where the stables once bustled are two facing rows of nineteenth century cottages, now modernised by a housing trust to provide small-scale accomodation. Not an echo remains of the horses' hooves, the shouts of the coachmen, or the laughter of Nell Gwynne.

The Bull.

The Bull shares with the defunct George and Bear Inns the necessary typical arrangement of a coaching inn - that is, an archway leading into a yard once lined with stables, and in turn emerging into a side street. The George Yard still comes out by way of Lavington Lane into Sheep Street, past the old Court House and Tanfield's House. Bear Court, the yard of the former Bear Inn, until quite recently emerged into Priory Lane by the Primary School, where garages now stand. The yard of the Bull emerges into Witney Street. The usefulness of this arrangement is obvious.

How far back the Bull Inn goes we cannot say, for no documents inform us, but there is every reason to believe that it goes back to 1500 and it may be as old as the George. As to the sign, it comes from that embodiment of animal vigour familiar to a farming community. The suggestion that it may derive from a papal bull or bulla is nonsense; if it had been understood in that sense it would never have survived the Reformation century, let alone the Commonwealth. The wealthy family of Hannes flourished in Burford from around 1500, and they were vintners, which marks them firmly as keeping the top level of inns. In the manor survey of 1552 John Hannes (Hanns or Hannys) holds one burgage tenement and a half, with a garden and a close of two acres adjoining. This can be identified with the three bay property of one-and-a-half burgage size described in a conveyance of 1607 as "*the capital messuage on the east side of the High Street over against the High Cross, known as the Bull*" and then occupied by John Silvester, inn keeper. (The High Cross stood in the road in front of the Tolsey.) This is not the present Bull property, which belonged to trustees for the Parish Church and was not sold until the nineteenth century, but the fifteenth or sixteenth century building three doors up the hill, with triple gables and carved barge boards, now housing Castles the butcher and its neighbour. The Hannes family went out of innkeeping in Burford, and John Silvester their tenant moved three doors down the street and, taking the sign of the Bull with him, set it up in 1610 where it has remained ever since. The Silvesters, like the Hannes, numbered vintners in the family. Robert Silvester of the Burford family was in 1597 made a freeman of the Vintners' Company in the City of London. The tombs of the Silvesters are in the Lady Chapel of Burford Church, a series beginning in 1568 and ending with a memorial in this century. Sadly, though the burial of John Silvester of the Bull is recorded in 1618, he does not have a monument there.

The arrival of the sign of the Bull was not the beginning of building on that site. Burford High Street was laid out for building plots in the twelfth century and to the best of our knowledge most of these were occupied before the middle of the thirteenth century. Buildings were altered and adapted but hardly ever cleared quite away, and most High Street buildings must include somewhere in their structure work of the mediaeval period, occasionally betrayed by the mouldings of a window or a filled-in door, but often not distinctive and therefore unidentified. The cottage belonging to the Bull and just behind it in Witney Street contains an arcade of early fourteenth century arches. It had until the late nineteenth century also a little lantern chimney, with tracery openings and a crocketed spire, a miniature of gothic detail. Work of the same period survives in Crypt House, a little further up the street. There was certainly a house on the site of the present Bull, and a distinguished one, for it was in important ownership. How much survives in the present structure we cannot tell. A handful of wealthy families dominated Burford at the close of the Middle Ages, their names occurring time and time again in the records of the properties. The Pinnocks were one of these families. In his will of 1473 John Pinnock the elder left the property that is now the Bull to his son John Pinnock the younger, who already owned the Angel next door on the corner of Witney Street, and then to his grand-daughter Margery and after her to trustees for the Parish Church. Mindful as men then were of the welfare of their souls, it carried with it an obligation to say prayers for his eternal well-being. In 1489 Margery, now married to John Petur of Minster Lovell, carried out her grandfather's intention, and made the property over. It was then leased out. We have the name of some sixteenth century tenants: the

Dalbys, father and son, burgesses and bailiffs of the guild, butchers in the middle of the century who also leased the shop then newly built against the back of the Tolsey, and one of whom was among the small group who set their names to the foundation deed of the Grammar School; and Andrew Ward, who acquired the lease by marrying the widow of Richard Dalby.

The first John Silvester may have acquired the lease from Andrew Ward, or he may have come here as sub-tenant. About 1620 one of those periodic rebuildings took place from which we may see the Bull emerging with all the details of an inn, though not with the brick front that i͏̸ ͏̸ ͏̸ today. In 1630 the second John Silvester obtained a furthe͏̸ ͏̸ ͏̸roperty in his own name, though now one John Cooke͏̸ ͏̸ ͏̸it for him. The rent had been 41s. a year; it was now £6 ͏̸ ͏̸paid to the churchwardens, but there was also a charge to ͏̸ ͏̸quer. At the Reformation, all chantry chapels were suppr ͏̸ ͏̸all bequests which required prayers for the dead were take͏̸ ͏̸ ͏̸rown. It was held that 10s. a year was due from the Bull on ͏̸ ͏̸ınt and in 1644 the Court of Exchequer noted that that it͏̸ ͏̸ ͏̸ years in arrears. In the 1650s Edmund Hemming repl͏̸ ͏̸ter. He took a new lease in 1658 to start in 1661 for £14 a y͏̸ ͏̸ıt improvement in the premises does this represent? Edm͏̸ ͏̸emming's widow Elinor took a further lease on the premises in 167͏̸ ͏̸ıen in 1697 Robert Aston, whom we have already met at the George moved over to the Bull. He died the following year and his impressive tomb in the churchyard bears witness to the status of the keeper of an important inn in a flourishing market town. He was not old at his death - a mere youth of 62 - but his tomb records a fact remarkable in that age of high mortality; all his ten children followed him at his funeral. One of them followed him at the Bull, for he was succeeded by his son-in-law Henry Tash, the husband of his youngest daughter Anne.

The Bull prospered. Just around the corner in Witney Street, adjoining the Bull back gate, was the **Talbot**. In the 1660s it had been the possession of John Payton, burgess and bailiff, a prosperous clothier. A clothier was not then what he is now, a seller of clothes, but a merchant who financed and organised the important cloth trade, from fleece to finished fabric, through the hands of all the various crafts who worked toward the final product. John Payton was one of the small number of Burford merchants and innkeepers we know to have issued tokens. Payton's tokens have a crouching Talbot - a large spotted dog - with *Iohn Payton clothyer* on one side and *In Burford 1666* on the other. There is another Payton token known, lettered *Iohn Payton - his halfpeny: of Burford 1669*. Only a very prosperous man would issue tokens, and the Talbot must have been a small part only of Payton's business. It had no future on its own as an inn. Already, before 1707, the large garden and stable of the Talbot had been taken into the Bull, and in that year the Bull absorbed the Talbot itself, Tash paying an extra £5 a year for it.

The Tashes were well connected in the trade. One of the legatees in the will of John Tash, innkeeper, buried here in 1747 was his cousin, William Tash, Esquire, Wine Merchant of the City of London. Towards the middle of the century the Tashes left the Bull and Edward Chavasse took over. His father Claude had arrived in Burford from France in 1710 probably in some Jacobite connection and founded here a remarkable family which not only played an important part in town life for over a century, but gave the nation two bishops, doctors, soldiers and an explorer. In Burford they were apothecaries, surgeons, innkeepers and schoolmasters, and Edward Chavasse of the Bull married Mary Castle, a relative of Dr. Castle of the Great House in Witney Street.

About this time we first see the functions of an inn mirrored in the pages of the county newspaper. Jackson's Oxford Journal began in 1753, just in time for the notorious disputed election the next year. Its weekly four pages reported national and international events as they came to hand, and local events and disasters as they happened, with court reports and details of executions. Half of the paper was taken up with advertisements and announcements; meetings of turnpike trustees, sales of property, auctions, plays presented by travelling companies, balls and assemblies. The great room - the later Market Room - at the Bull was in constant demand. In 1758 Burford suffered a terrible epidemic of small-pox, and over two hundred persons of the town were buried in three summer months. But in September we are informed that the Subscription Assembly and Ball in the Bull Great Room will be continued after the epidemic. The races were a great attraction that brought the fashionable world to Burford and from the Restoration to the Regency the town possessed something of the character of a resort. Edward Chavasse was succeeded by Edmund Fellowes in 1768 and he announced a Ball to mark the Races but that was only one event in a full social calendar.

The Bull is unique for in the whole length of Burford High Street it has the only brick frontage. In the sixteenth century it was the vogue to put half-timbered fronts even on to stone buildings. In the eighteenth and nineteenth centuries it was fashionable to put sashed and parapet fronts on to earlier gables, often not matching the interior arrangements of the buildings. But only the Bull has a Georgian brick front and it must be a deliberate exercise in promoting the image of fashion. Exactly when it was done is not known. There was a sudden jump in the annual rent from £6 to £14 in 1661 which must reflect a great remodelling of the Jacobean inn. Then, a century later, the churchwardens, the landlords, spent £196 on the Bull at the end of Chavasse's tenure, borrowing £41 from one of their number to do so. For this sum a serious rebuild would have been possible. The payments in the accounts for this work began in October 1768 and lasted until March 1770, but the majority of the thirty-one separate entries give only the name of the payee, and there is no tell-tale entry for bricks. John Humphreys was paid 4s 6d for digging nine loads of stone; Edward Ansell's bill for *lath, lime and hair* came to £14 12s 8d. The work for which John and Richard Osman were paid £13 17s 2d was simply *Great work measured* and £14 4s 6d for *Day bill, materials and labourers*. £18 5s 6d was paid to John Simpson without detail, and most of the other entries are the same.

In July 1770 *"the Balls held during the races at Burford will be at Mr. Fellowes, at the Bull Inn, on Monday and Thursday evenings, the 23rd and 26th of July"*. Edmund Fellowes did not stay long and was replaced by John Baldwin in 1774. In December 1775 a Card Assembly and Ball is announced for January; Admission 3s. An expensive evening indeed for the gentry when a labourer with family would have to be content with 10s a week. In 1782 there was an "undress (i.e. informal) Ball" at the Bull for the second evening of the races.

Travelling companies of players presented their offerings - all presumably in the Market Room. In July 1780 Mr. Richardson and his company *"return their most grateful Thanks to the Nobility and Gentry for the Many repeated Marks of their Favours this and many past seasons at Burford .... now advertising plays in Chipping Norton also: the Tragedy of Percy ... the justly admired comedy The School for Scandal ..... the tragedy of Hamlet, with the Entertainment of The Ghost"*.

The Bull Inn, Burford's only brick front.

There was a rougher sporting side: cock fighting took place in a cock-pit at the Bull as well as at the less fashionable inns:

The Annual Subscription Cock Match will be fought at the Bull Inn in Burford, on Wednesday the 17th Day of March instant, 1784. Subscribers and others that chuse to send in a Cock, or Cocks, are desired to send them in on or before the 8th of this instant March. - The Cocks to fight for five shillings a cock. - To begin fighting at Ten O'Clock, and no Persons admitted but Subscribers, or those that Dine. - Dinner on the Table at Two.

The rustic side of life surfaces as well. In 1765 a farmer in pursuit of a bet promised marriage to a servant girl at the Bull, but then cheated her of the promise by failing to provide licence or ring. He was treated to the bucolic custom of "rough music" - a traditional cacophony of beaten utensils and ribald jeering.

In 1785 the holder of the Bull was changing again:

The Bull Inn, Burford. To be lett and entered upon at Lady Day next, most advantageously situated in the High Street, now in the occupation of Mr. John Baldwin. The above Inn has many years maintained a capital run of Business, for which it is accommodated with every Convenience. Particulars may be had of Mr. James Merywether, Attorney, at Burford, or of Mr. John Patrick who will shew the Premises. N.B. Immediate possession may be had, if desired.

What was the immediate result of the advertisement we do not know. Not until January 1790 do we read that Mr. Stevens has taken the Bull. Mr. Stevens had been the butler at Cliveden, and aimed for the height of fashion. The rate of advertisements in the Journal is much increased. There was an Ordinary for each day of the Summer Races. (An Ordinary was a regular meal provided at a fixed price; the term has gone out of use, and its place has only partly been filled by *Table d'Hote*.) Some agreement had been reached between the Bull and the George, for when the Ordinary was at the Bull on the first and third days of the meeting it was at the George on the second and vice versa. Balls were held at the Bull on the second day of the meeting. Meanwhile the regular social life of the town continued in 1792:

Burford Card and Dancing Assembly will be held at the Bull Inn, Burford on Thursday the First of November & will be continued monthly through the winter.

From time to time there is an important addition: *There will be a full moon*. The flood lighting of Burford's High Street these days makes it difficult for us to realise the darkness of the past or the roughness of pitched stone pavements. 1793 was a good year at the Bull. In July there was a Ball following the annual Venison Feast:

Burford Annual Venison Feast with an additional Buck will be held at the Town Hall on Tuesday July 30th 1793. Dinner on the Table at One O'Clock. N.B. A Ball at the Bull in the Evening.

The Town Hall refers to the upper floor of the Old Grammar School building on Church Green. The Venison Feast went back to the mediaeval right of an annual town hunt in Wychwood commuted in Elizabethan times to the gift of a buck from the Forest.

In December Mr. Clark, dancing master ... *presents Compliments to the Publick & informs them that his Ball for the young Ladies educated by the Miss Pritchards will be held at the Bull Inn on Tuesday next, the 10th instant, when the following Dances will be performed: the French Minuet, with the new Graces and the much admired Minuet Deiga, with Cotillon and Scotch Steps only; Minuet de la Cour, with the Gavott; Cotillons in the present Taste, Country Dances etc. After his Pupils the Ball will be publick for those Gentlemen and Ladies who will favour him with their Company. Tickets 3s 6d each, Tea and Coffee included.*

The Miss Pritchards conducted their school in a house long ago demolished which stood by the northwest of Church Green, where the school's Lenthall House garden is now.

In 1800 the George was closed, and the triumph of the Bull was complete. In January of 1801 the Bull advertised:

Burford First Ball. Cards and Tea in the Bull New Room, the Union, will be held on 27th January Inst. (the Tuesday before the Full Moon) with the Harp and Violin.

Ladies' Tickets 3s 6d. Gentlemens' 5s 0d.

The season closed with a last assembly on March 24th. The prices were the same. Bearing in mind the general value of money at that time, the Burford season could be very expensive for a family man.

In 1797 the Journal reported a disaster at the Bull. A candle left by an ostler in the stables had fallen on to straw, an ever present danger in the days before electric light, and before the alarm was raised the building was burning strongly. Seven coach horses were so badly burned that they died within a few hours, and before engines arrived from Witney and Bampton the building was totally destroyed. The Journal commented on the absence of an engine in Burford at that time.

The races must have brought many dubious characters to Burford and not a little scandal from time to time. In 1803 one man's troubles reached the national press. Mr. Arthur Barry Shears, an Irish gentleman, arrived with his wife at the Bull Inn from Cheltenham, where he had passed all the previous year "*in a pleasant circle of acquaintances*". At 10 p.m. the lady retired to rest and Mr. Shears who had previously appeared to be in the best of spirits - "*of light but inoffensive manners*" - produced a large pistol, aimed at his head and lodged a quantity of grape shot in his brain. He lingered until the following afternoon. The Gentleman's Magazine indicates pecuniary embarrassment, but also mentions that two of his brothers had been executed for high treason during a rebellion in Ireland.

* * * * * * *

Great Coaching Days.

Inns are inseparably linked in the popular mind with coaches, Charles Dickens and deep snow at Christmas. The reason is a coincidence in time. The steady growth of trade and business through the centuries meant that travel grew as well. The net work of turnpike roads and stage coach routes developed rapidly in the late eighteenth century and reached its peak in the first three decades of the nineteenth. Charles Dickens began to publish *The Pickwick Papers* in which coaches, inns and Christmas repeatedly appear, in the year 1836; and the whole period coincided with a small "ice age" when winter temperatures were low. Inns were a vital part of the transport system: they provided the stopping points for picking up and setting down passengers, and the only physical comforts in what was in spite of all efforts an arduous and uncertain method of travel. They also supplied the necessary changes of horses. Burford was splendidly situated to be a coaching centre. In the late seventeenth century Ogilby's maps had placed Burford on two important routes: from Salisbury to Chipping Campden by way of the Thames crossing at Radcot Bridge, and from Bristol to Banbury through Cirencester. Because of the Thames marshes west of Oxford the London and South Wales route passed south of Burford through Fairford, but such improvements as Seven Bridges Road (Botley Road) in Oxford and the Swinford Bridge brought the route through Burford and Northleach. In the late eighteenth century the Oxford Journal frequently carried items remarking on the increase of traffic along the improving turnpike roads. Cheltenham grew rapidly after George III spent five weeks there in 1788; it became a fashionable magnet and the population rose from 3000 in 1801 to 23000 in 1831. From Oxford to Burford is about twenty miles - a fair stage for a change of horses, and a convenient place before starting on the crossing of the Cotswolds with the cruel gradients and the winter hazards. The Bull kept a wall of book-shelves in the coffee room filled with three-volume novels for the entertainment of travellers detained in Burford by mishap or deep snow. In 1772 the coach from Gloucester to Oxford breakfasted at Frog Mill and dined at the Bull at Burford. Returning it breakfasted at Burford and dined at Frog Mill. At Oxford it connected with coaches to London. In the same year the Stroudwater Coach from Stroud to London via Cirencester, passed through Burford. Inside passengers paid one guinea each, and were allowed 14lbs. of luggage, with excess baggage at three halfpence per lb. But travel was improving. The next year the Burford, Witney, Oxford and Thame Fly began, leaving the Bull every Monday, Wednesday and Friday morning at 3.0 a.m. breakfasting at the Cross in Oxford, dining in Stanmore, and arriving in London the same evening. Burford to London 15s; outside passengers half price. 20lbs of luggage allowed, and excess at 1d per lb. The demand to complete journeys within twenty four hours, and the relative slowness (to the modern mind) meant far more travel during the hours of darkness. In 1773 an ostler of the Bull Inn waiting at its back door for the Gloucester Stage to arrive at 2.0 a.m. was witness to a woman falling from an upper window in Witney Street. (The story is intriguing: she had been bed-ridden for ten years, but on being taken up in the street with only a dislocated shoulder and bruising, found she could walk again.) The east-west traffic did not then as now pass by above the town but clattered and jingled along Witney Street and Sheep Street.

The great days of coaching came in the next century, when at the height of the period Burford is said to have had forty scheduled coaches stopping here in each twenty-four hours. The more prestigious coaches were given names to mark their status. The Berkeley Hunt came through from London in the middle of the afternoon every day except Sunday heading for Cheltenham, at some point passing the return coach which had left Cheltenham at 6 a.m. The Retaliator ran two hours behind on the same route. The Champion came through shortly before midnight heading west; the east bound Champion came through in the afternoon. The Veteran headed west through Burford in the late afternoon. The Royal Mail from London came through at 4 in the morning. The York House Coach from Oxford to Bath via Burford and Cirencester went through west bound on Monday, Wednesday and Friday mornings and returned on the alternate days. When one considers the timings that were kept and the state of the roads it is no wonder that the drivers acquired heroic reputations or that it was a frequent occurence for the horses to drop dead in their traces. The most trying times were at the end of the month when heavy loads of the growing magazine trade were carried for fast delivery. Often quite unacceptable risks were taken and the Oxford Journal carried many reports of hideous accidents. Those named above are but a few of the many scheduled coaches. In addition there were the slower stage wagons, principally for goods but carrying passengers as well if they were poor or not too particular. Then there were the local carriers, plying from lesser inns. Ostlers were on hand throughout the twenty-four hours. Post boys hung around the yards waiting to take messages or parcels. The streets can seldom have been silent day or night.

Inns were not only the stopping places for coaches, but they themselves provided transport at need. Travel was very expensive: the hire of a chaise and pair cost a shilling a mile and innkeepers complained that at that they were making a loss. In 1802 a general meeting of innkeepers in the area from Bath to Worcester to London placed an announce- ment in the Journal setting out in detail how at 12d per mile they were not covering their expenses and warning that the rate would rise to 14d per mile for each pair of horses and 6d per mile for a saddle horse. It makes transport to-day look cheap.

* * * * * * *

Mr. John Stevens of the Bull died in 1820 and his widow carried on, followed by her son. Burford continued to be a fashionable resort. On August 25th, 1832 the Journal reported:

"*About 3 o'clock Sir Henry Peyton's handsome equipage brought down a party of Noblemen and Gentlemen ... to the Bull Inn ... to an annual Venison dinner ... after partaking of an elegant repast they left Burford about 6 o'clock. Another party of Gentlemen from Oxford and the neighbourhood also dined there on Monday, in addition to the numerous venison feasts already had this season, and yet to come at various inns in the town.*"

But in October that year young Mr. Stevens left the Bull to take the Ram at Cirencester. Perhaps he had a premonition of what was to come, for relatively swiftly the roads of England changed and with them the inns. In 1837 the first railway line ran out of London; in 1850 the last scheduled stage coach ran through Burford. All the schemes from 1840 until 1900 failed to provide a railway for the town. Its communications withdrawn, the open downs enclosed and the races moved away, the place lost the character of a resort and was driven back into its role as a market town. Agriculture was still prosperous and the town crowded for fairs and the weekly market. Again the Bull provided a Market Ordinary for farmers and supplied a more stolid and less fashionable society. Then about 1870 the great depression in farming that was to last till 1940 set in, the market failed and Burford became quiet indeed. The Irish John Dunphy kept the Bull after Stevens, then Jabez Appletree, then Robert Woods, Humphrey Porter and after him his widow span the years from 1860 to 1885, Orlando Woodward, Henry Gillett and then at the end of the century Frank Arthurs. He is said to have been sitting late with a glass one warm evening by an open window when a circus travelling by night, as was the custom, paraded through. The trunk of an elephant snaking through the window and sweeping the glass from his hand scared him into sobriety.

Now we are almost into living memory and a familiar world. In 1899 the Bull advertised "*commercial and family stabling, Cyclists Touring Club facilities, billiards*". In 1907 there was a motor garage, and fishing was offered. At first visitors came for the remoteness and quietness;

then touring became a popular recreation and Burford was back in the busy world.

In the small hours of one night late in July 1982 Burford was awakened by the shouts of men unrolling hoses down the High Street to the river and the steady pulsing of pumps. A passing driver had spotted smoke curling from the roof of the Bull. The fire had already obtained a serious hold and the staff escaped with difficulty. Eleven engines attended the blaze and only a high turntable ladder from Oxford pouring water into the building from above mastered the blaze as the early light spread in the sky. The dawn showed a sorry sight; the windows were gaping, the roof had gone. The fire had spread to the next house up the hill, but prompt action had saved the mediaeval house beyond that. Over the next few weeks the facade was brought down and then slowly the interior and the front were carefully reconstructed. In the summer of 1983 the Bull was back in business.

The Bear Inn.

The third Burford inn to possess the characteristic plan of an archway on to the High Street leading to a yard emerging into a side street is the Bear, no longer an inn but now Bear Court, a pleasant development of shops and in the summer a courtyard of flowers. As an inn the Bear must always have been rather a disappointment. The Bull and the George both lie near the market centre and the crossing of the roads. The Bear, a late comer, is too far down the High Street to attract travellers or much trade. The corner of the High Street and Priory Lane is dominated by the enigmatic Falkland Hall, a twentieth century name for which I can find no warrant. It was once a great house built by the wealthy cloth merchant Edmund Silvester for his own occupation in the year of Queen Elizabeth's accession. Next to that and wrapping round behind it, the Bear Inn was built sometime in the seventeenth century. It seems very likely that it was the work of the Matthews family who appear in Burford at the time of the Civil War and may have hoped to profit from the new importance of the Priory a short distance away. The owner of the Priory was William Lenthall, Speaker of the Long Parliament, who had outfaced King Charles.

Briefly, between the King's execution and the proclam- ation of the Commonwealth he was Head of State for England. There is no mention of the Bear before then, and for the next thirty years Mr. Thomas Matthews of the Bear was an influential man in Burford. His children appear in the Baptismal Register from 1648 to 1656. He was a burgess and four times between 1655 and 1675 he was one of the bailiffs of the town. He was an important tenant of the Lenthalls, Lords of the Manor, renting large new inclosures of farming land. Some of the farthing tokens he issued from the Bear survive: *Thomas Mathewes At The Beare* around his initials on one side and on the other a chained bear. William Lenthall was driven into private life by the Restoration of Charles II and he died in 1662. His son Sir John, Baronet of the Commonwealth and a knight under Charles II, was a none too scrupulous courtier - the contemporary gossip Anthony Wood calls him *"the great liar and braggadocio of the age"* - but he and the Burford Races may have brought trade to the environs of the Priory.

The success of the Matthews did not last. Thomas Matthews' wife Elizabeth died in 1671 and he himself in 1680. His son Jonathan was baptised in 1648, married in 1675 and died four years after his father. The Vicar John Thorpe wrote in a note that year ... *"the mathews who lived at the Beare Inn in Burford are extinct & the farme is disposed elsewhere"*. Mrs. Ann Matthews, widow of Jonathan, lived on in a house in Priory Lane for some years but the Bear is hardly heard of again. Possibly it flourished at fair times when the length of the High Street was bustling, or the cockpit behind it brought trade. Just one entry in the Oxford Journal shows that it was still in business: on July 28th 1760 the Land Tax Commissioners met at the Bear. There must have been some good reason why on that occasion it was not at the Bull or the George.

The Bear is absent from the lists of victuallers from around 1774 to after 1800. When next we meet the name, it is the property of the Lord of the Manor, now William John Lenthall. Mounting debts largely caused by the expense of the enclosure of the parish drove the Lenthalls from Burford Priory in 1828 and many Burford properties were sold off separately at that time. In 1830 John Jones, carpenter and builder, raised £300 to buy from the Lenthalls the block of property on the corner of the High Street and Priory Lane, including

The yard of the Bear Inn as it was.

the Falkland Hall and the Old Bear "*formerly used as an inn or public house but for some time past divided into private apartments*". The Bear and the Falkland Hall had been lumped together and shared the common fate of large redundant buildings, being carved into tenements. Already George Akers, carpenter, was established in a workshop somewhere here. John Jones, builder, saw the usefulness of the yard for his business. About 1840 he also re-opened the old inn as a public house and so it continued until recent times but generally kept as an adjunct to another trade. After Jones came Emanuel Sperinck, then John Butler, carrier, Frank Arthurs, hurdle maker (later of the Bull) and two more before it came into the hands of the Hall family, blacksmiths at the Forge (now Quill Books) three doors nearer the bridge. With them it remained until it closed in the 1960s to re-open as Bear Court. Billy Hall went on sitting outside the Forge on sunny days until time removed him and the Forge as well was changed.

The great coaching inns such as the Bull and the George and their less successful competitor the Bear demand specialised premises.

then we shall walk up the High Street to the crest of the hill, looking as we go.

The Swan in Cobb Hall.

The first **Swan** opened near the bridge in Cobb Hall. This was a large courtyard house, Tudor or earlier in date, and only the archway and two rooms, reduced to a single storey, survive above ground, where the outline of three filled-in windows can still be seen on the High Street; though, as the owner of what is left, I can confirm that the pitching of the courtyard is still in position two feet below the present gravel. It was left to the town by George Symons in 1590, the rent to be used for the poor. By 1600 it was in business as the Swan. In 1622 there is an entry in the Parish Register of the burial of Thomas Hushe, "*stabbed at the Swanne*". It was an appropriate enough name for there are always swans nearby on the river. Richard Norgrove, innholder, took a lease of it in 1629 for 21 years at £8 a year, when the rent of the Bull was only £6. His will was proved in 1636 and the attached inventory indicates a large and successful business. It lists a parlour, chamber over the parlour, the hall, chamber over the hall, chamber over the hall next the bridge, the forestreet chamber, the little chamber, the buttery chamber, the great chamber, the kitchen, the stable in the court, the beer cellar, the wine cellar, and has long separate inventories for linen and brass and pewter. His total valuation was £208 12s 3d, comparable with the valuation of the keepers of the George at the time. The inventory is much like those at the George. Here is one room:

Item in the chamber over the p(ar)lour
item one joyned beadsteed and
    trundlebed wth cords & matts   xxx s
item curtaynes & valens & curtayne roodes   x s
item two coverleds & blanklets   xxxiiij s
item two featherbeds one fether bolster,
    one other feather bolster   5 li 18 s 10 d
item two flocke bolsters   v s viij d
item one carpett vj cushins   ix s
item one chaire & fower stooles   vij s
item one pr of doggs bellowes & fier shovle   ij s vi d

item one pr of doggs bellowes & fier shovle   ij s vi d
item one side cupboard & carpett   v s
        Som   xj li ij s

It was leased again in 1648 at the same rent to Edmond Hemming, barber surgeon, later keeper of the Bull, for his own occupation. He assigned his lease of the property "*with a signpost and sign of the Swanne standing at the door thereof*" in 1650 to Robert Collier of Taynton, whose tenant here was Richard Willett, later of the Catherine Wheel, who died in 1678. The Willetts, too, had links with pharmacy: Nicholas Willett, apothecary, took the Crown in 1734. There seems to have been a link between medicine and inn-keeping at this period. In the Burgage rent roll of 1652 it is inexplicably (since Symons had left it to the town in 1590) headed "*The Land late the Lord ffalklands Land*" and described "*A tenement lyinge att the Lower end of the towne on the westside thereof called the Swan now in the tenure of Richard Willett and now towne land*". In the later roll of 1685 it is "*heretofore called the Swan*". It had ceased to be an inn by 1679 when Paul Silvester rented it, and he and William Rogers, who renewed his lease in 1687, and the Flexney family who followed them in the next century, were all prosperous clothiers, merchants organising the cloth trade. The name of the Swan, however, lingered.

In the nineteenth century it suffered from the decline of the town and was turned to other purposes. Part of it became the elementary boys' school, part was a joiner's shop, but all became dilapidated. In the great auction of charity properties it and the three cottages in the close behind were put up for sale in 1860, and again in 1862, but found no takers. It was subsequently sold for £200, and was largely demolished about 1876, and the property was joined to the former Vicarage. All that remains is now part of Cobb House.

We have looked so far at large or purpose built premises housing inns with lives measured in centuries, and well documented. Other inns or ale-houses had shorter lives, in properties that moved in or out of the trade, and left little mark. The lists of inn-keepers' recognizances that formed part of the licensing system have survived for the second half of the eighteenth century, and from 1774 the names of the inns are recorded together with their keepers. Many of these men are known in other records by their day-time occupations. The lists, in spite of their administrative context, are often only approximate in their recording of names of persons or premises, and with certain names disappearing only to appear again. I have reservations about their total comprehensiveness, but they must provide a fair record of the inns in all Burford at the time.

1774

Black Horse	Ranchford Strafford	Quart Pott	Thomas Legg
Board and Letters	Robert Strafford	Red Horse	Ephraim Sperinck
Bull	John Baldwin	Red Lyon	Ann Hanks
Chequers	James Wickens	Rose and Crown	Ann White
Cooper's Addice(Adze)	John Kempster	Sun	John Brown
Cross Keys	William Strafford	Swan	John Day
Dolphin	Jane Wells	Three Oranges	Richard Winfield
Eight Bells	Isaac Coburn	Three Towers	George Cooper
George	Thomas Clare	Waggon & Horses	Robert Osman
Greyhound	Thomas Buckland	Wheatsheaf	Jonah Freeman
Horse and Jockey	Edward Wallis	White Hart	Jacob James
King's Arms	William Boulter		
King's Head	John Eldridge	*Upton and Signet:*	
Lamb	Thomas Merrick	Bird in Hand	William Eden
Mermaid	Christopher Kempster	Lamb	Charles Maisey
Pound of Candles	William Hulls	Rose and Crown	Joseph Strafford

By 1787 the Board and Letters, Chequers, Cross Keys, King's Head, Pound of Candles, Three Oranges, Waggon and Horses, White Hart and Bird in Hand have ceased to appear, but the Old Wheatsheaf (James Wiggins), Old Mermaid (William Sessions), White Horse (Joseph Strafford), Royal Oak (James Strafford), Bell (Alice Nunney), Glaziers Arms (John Beal), Unicorn (Samuel Heath) and Fleece (Henry Buckland) have arrived.

In 1800 the Bird in Hand is back, but the Black Horse, Adze, Horse and Jockey, Royal Oak, Glaziers Arms and Unicorn have gone.

I have the following list for 1854 from another source: Bear, Bird in Hand, Birds Nest, Bull, Coach and Horses, Eight Bells, Golden Ball, Golden Fleece, Greyhound, Kings Arms, Kings Head, Lamb, Masons Arms, Mermaid, New Inn, Plough, Red Lion, Rose and Crown, Rose and Crown (Upton), Royal Oak, Swan, Three Pigeons, Wheatsheaf, White Hart, White Horse. Mr. Fisher in 1861 reckoned that there were 16 public houses, but 19 in the previous year.

The majority of these, but not of course all, were in the High Street. It appears from these lists that the general estimate of between twenty and thirty establishments in business in Burford, around one for every sixty inhabitants, from Tudor times to the nineteenth century (when numbers were falling) is correct. This agrees with figures from other market towns. Of course, then as now, the town served more persons than lived in it. It also appears that while many houses represented the part-time occupations of families whose main income was from building trades, crafts or cartage, many were also businesses such as saddlery or tailoring, where work was done on the premises. Some also were shops, especially provision shops of various kinds.

An example of a provision or victualling shop of some kind is the **Unicorn,** which appears in 1787, kept by Samuel Heath. He is listed in a directory of 1792 simply as Maltster and Shopkeeper, marked as a Freeholder. The Journal carried a notice of his bankruptcy in 1799, with a sale of his effects at the George. The sale notices make it clear that the Unicorn was a sizeable property in the High Street, with malthouse and stabling. The **Plasterers Arms** (which does not appear in the above lists, being sold in 1765 and going out of business, "formerly kept by Stephen Woods") and the **Glaziers Arms,** in occupation of John Beal and also sold in 1799, are obvious examples of building trade connections.

A Walk up the High Street.

The **King's Arms** is a relatively common name that may have been used by several different establishments. There was a burial in 1695 of *"an old travailing beggar man died at the Kings-armes"* but this may not be the same as any later inn. The most successful King's Arms was on the south corner of Lawrence Lane and the High Street, on ground that had once belonged to Brasenose College. It was later the Hopewell House of Mr. Potter, corn merchant and prominent Methodist. When Mr Potter's widow died in 1933, the property was purchased for the Grammar School. It became first the Girls' Department and then, when in 1958 the new school buildings were opened at the top of the hill, the Junior Department, and later still when boarding for girls was introduced, the girls' dormitory accommodation of Burford School's Lenthall House. Twentyfive years ago the matchboard partitions of rooms in the roof which must once have had dormer windows were still in position. This must have been the King's Arms kept by William Boulter and found in 1773 in the Oxford Journal and appearing regularly thereafter. Auctions were often held there, and from 1799 until 1803 there was a Ball held to follow the Venison Feast. In February 1845 seventy gentlemen dined there on a steeple chase day, when the racing had been cancelled because of the weather, but that same year the contents were sold up, Mr. Ford retiring. There is one further directory entry in 1852. The stores and stables of the inn extended back down Lawrence Lane and were used by Mr. Potter for his corn merchant's business. Perhaps the ballroom had already come down in the world to provide the Murdoch weaving sheds, but these in turn were demolished to make way for the New Assembly Hall, now the dining room of the school's Lenthall House.

The **King's Head** was less significant but more of a problem. There was a King's Head on the south at the end of Witney Street, called "former" in 1716, but the later one of the 1840s, kept by William Smith who was also a baker, may have been in what is now Archway Cottage, with its mounting block, near the bridge.

Why a town as far from the sea as Burford should have such an exotic creature as a mermaid on an inn sign I cannot tell, but perhaps that century when all Englishmen first surmised that the sea was in their blood brought her inland. In 1706 the first **Mermaid** occurs, on the west side of the High Street and kept by Joseph Overbury. From the deed in which this reference is found all we learn is that it is not on the north corner of a block, where the later Mermaid stood. Overbury's will proved in 1718 is unrevealing. After Overbury it was kept by Simon Badger, mason (builder) and church-warden, whose son was baptised Thomas Overbury Badger. The Bailiffs of the Corporation were frequently there: in 1733/34 they *"spent at ye Mermaid 0. 0. 8d"*, and it was well enough regarded to be a place for auctions. At the end of the century two Mermaids co-existed: in 1782 the **Old Mermaid** was up for letting and enquiries were invited of Mr. Benjamin Swancott upon the premises. The sign moved before 1800 when the Lower Mermaid in the High Street was again to let, as Mr. William Wiggins

37

was going into another line of business. It found a taker and Richard Taylor was there by 1819. Now we know exactly where this Mermaid was, for in the first Dr. Cheatle's records of 1819, Richard Taylor is indexed as "*of the Mermaid*", but his account is made out to Mr. Taylor "*at the Pump*". That pump, the public pump for the lower High Street, stood on the corner of Priory Lane six feet or so from the side of the present Cotswold Arms, and it is shown in Buckler's drawing of the site in 1821. The Taylors stayed, Richard being succeeded by Thomas and then John and then Mrs. Sarah Taylor. The family combined keeping the inn with the trade of tinsmith and brazier. Just how much of a sideline the keeping of what was now a public house could be is plain in the censuses where simply the occupations are given and the public house ignored. Joseph Brunsden, slater and plasterer, was there in 1876. In the first large scale Ordnance Map of 1881 we see that the present Cotswold Arms was indeed the Mermaid, but then the name and the keeper change. Bartholomew Collingwood Fisher, carrier and general dealer, came in and the sign of the **Cotswold Arms** went up. Why he chose that name I cannot guess, nor what arms should signify the Cotswolds: sheep perhaps, quartered with the tools of the mason's trade. The Cotswold Arms it is still, with its low ceilings and small windows at least looking the traditional part.

That was not the last of the **Mermaid**. Just up the High Street where the seductive Mermaid hangs over the pavement now was an inn called the **Three Pigeons.** The name is not older than the early nineteenth century, however old the building, and the first keeper was James Dunphy in the 1820s, an Irishman who later kept the Bull. Fisher was the keeper in 1876, just before he moved to the "lower Mermaid" and changed the name of that. The Three Pigeons ceased trading for a while but after the last war, when the front of the property was rebuilt in a nondescript style but at least in stone, it re-opened as the Mermaid.

The early nineteenth century **Bell** was also in the High Street, on the west side, I suspect above the Tolsey. We have already met this name a few years later lower down on the east of the Hill. Perhaps the keeper moved and took the name with him. In 1818 a property on the west side of the High Street associated with the saddlery trade of the Jeffs family is described as "*known by the name or sign of the Bell*", but although the documentation continues to 1858, it is not called the Bell again, nor are any of the occupants described as innkeepers or victuallers. The **Bell** reappears in 1823 now on the east side of the High Street and kept by the Nunney family, who were in the plastering and slating business. It was a charity property, part of the Church Estate, and in 1859 in a Charity Commissioners' survey it was described as "*dirty and dilapidated*" and recommended for sale. The Nunneys bought it in 1860, but its last entry as a functioning public house had been in 1847. It was situated on the east side of the High Street and the census returns place it five households south of the Church Lane corner.

After the Mermaid we can hardly be surprised to find the **Dolphin,** in fact the beginning of a school of dolphins, for there was another Dolphin over the river in Fulbrook, vanished now but surviving in the name of the house that was built with its stones on Westhall Hill. The Burford Dolphin was next but one to the Methodist Church, formerly the Chapman mansion, in the southern part of Belinda's dress shop. Though the property is far older, it is first mentioned as the Dolphin in 1735 when it was kept by a Kempster. By 1778 it had become the **Plough** and was kept by Edward Allen but belonged to William Chapman, a man of fortune, who had come to Burford and bought the mansion in 1766 and steadily acquired the properties round. There were three generations of Chapmans, and the last died unmarried at the age of 79 in 1845. He had a constant man-servant, James Merchant, to whom he made over the Plough in 1826. The Plough, kept by the Rouse family and then by Edward Stiles, lasted until 1876 when it was bought by George Hambidge who owned the long established grocer's shop next on the south side and it ceased to be an inn.

The **Rose and Crown** (not the one at Upton) had a similar existence. Its studded door is on the High Street, with the Tudor Rose and Crown patterned in nails and the date 1915. Much older in the stone spandrel above, is the merchant's mark of Simon Wisdom, who dominated Burford in the reign of the first Elizabeth, and the date 1578. As an inn it lasted as far as we can tell from the end of the eighteenth century into this century, though it is not in the directories after 1918.

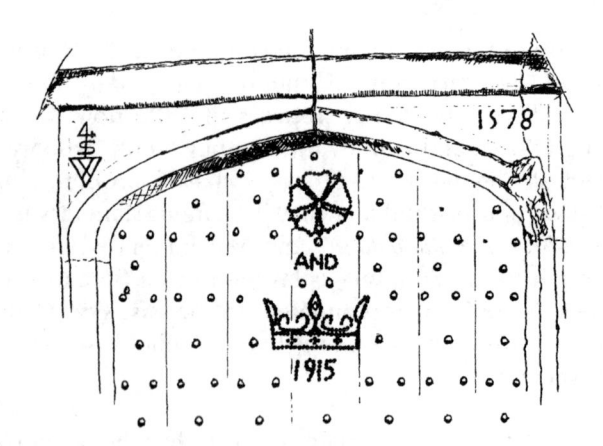

The doorway of the old Rose and Crown with Wisdom's mark above.

Other High Street properties have moved in and out of the trade, being inns and shops or private houses by turns. With no distinctive marks it is often difficult to place a departed public house precisely. But there is no doubt about the **Red Lion,** first mentioned in 1770 in the Oxford Journal. It had a sporting character, the keeper racing a horse on the downs and in 1789 organising a pigeon shoot and giving a silver cup as a prize. The inn disappears from the directories after 1864, was vacant in 1871 and became Drinkwater's butchers' shop before 1880. Then it was Barclays Bank until the 1970s and is the Red Lion book shop to-day, with nothing beyond the name to suggest outside that it was ever an inn.

Facing it is the **Golden Ball,** now renamed the **Golden Pheasant.** The Golden Ball is, of course, the sun, and there is no need to go all the way back to the "glorious Sun of York", King Edward IV for the origin of the sign, or to the emblems of Italian merchants. As an inn it is not so old. John Hunt operated a wine and spirit business there from the 1840s, and was followed by his widow Anne, and then by his son Frederick, when the business became Hunt, Edmunds. The name Golden Ball was in use in Sheep Street in 1758, and in the High Street in 1836, but in the directories the Hunt business is always described as Wholesale and Retail Wine and Spirit Merchants. The first use of the inn name I can trace for this property is in the 1920s, when it was kept by Edward Hunt. The building is old, but the fresh appearance of the stone work is due to a near catastrophe of the 1970s. During alterations some vital load bearing part of the front wall was disturbed and the entire facade began to move out into the street. All traffic was halted for a day while the front was shored up, but a total rebuild was needed, and although the old stone was used, it was redressed. The name was changed around 1980.

Inns became shops; some shops have become inns. On the corner of Witney Street is **Burford House,** briefly **Andrews Hotel,** but for decades before that the **Corner House**. This was a furniture store until between the wars when it was bought for restoration by Mr. Horniman, of the Priory. In the process of rebuilding, the corner was cut away into its present shape with the intention of improving visibility for traffic. It opened as a hotel and traded as the Corner House until its name too was changed by new owners in the 1980s.

Now as we pass the centre of the town where the old market Cross stood outside the Tolsey it is time to breathe a sigh for the High Street inns that have vanished leaving not even their names behind them. For some, such as the **Blackamoor's Head,** the **Three Sugar Loaves** and the **Three Cups,** we have names but no location. In 1691 the infant child of William Herbert, described as the keeper of the **Half Moon,** was buried. Richard Willett we have already met at the Swan, but at his burial in 1678 he was the keeper of the **Catherine Wheel.**

One sad romantic story must not be lost. On November 9th 1765 an advertisement appeared in the Oxford Journal. *"A lady enamoured of a University man asks him to call at the Green Dragon, Burford ... "* It may have been a transient inn, but if ever there was a Green Dragon at Burford, this is the the only reference to it. Sad lovelorn lady yearning for scholarly charms! Where had they met before? Did she not know his name? Was she possessed, like Dante, by an unforgettable sighting? Could she have mistaken the point on a coach journey where their brief encounter took place? There was a Green Dragon at Chipping Campden. Did they meet here, or only now in the Elysian Fields? Or is it a message in code?

Now we can go on up the hill. Castles, the long established butchers, occupies two parts of the splendid three-gabled building in which the Bull began and which, around 1620, was lived in by Richard

Merywether who succeeded Simon Wisdom as Alderman of Burford, and built an inn himself in Witney Street. The property was already regarded as two by 1731, though they were in the same ownership. The southern third was in business as the **Wheatsheaf** by 1758 when a creditors' meeting was advertised there, and it is called by that name in the deeds from 1800, when it was kept by Jonah Freeman and then by James Wickens. His family seems to have held it until around 1860, and soon afterwards it disappears from the records. For a while in the late eighteenth century it co-existed with an **Old Wheatsheaf**.

Once the Bull Inn, the uphill third part became the Wheatsheaf.

Next to the Wheatsheaf is the house in which Simon Wisdom lived in Tudor times, refaced by Richard Whitehall in eighteenth century style. Above that was, until recently, the **Highway Hotel**, another twentieth century creation. The building was already old when the first Elizabeth was Queen; two large rooms fronted on to the High Street with a passage between them, and there was a storey above with a roof parallel to the road. At some time another storey was added by carrying the front up into the roof. Then two wings were built out at the back; one at the south perhaps before 1600, and one at the north a century later. Time has veiled the name of the first builder, but from before 1600 the Hunt family lived here, trading as ironmongers, brass and coppersmiths, and pewterers. "*William Huntt iremonger*" who was buried in 1613, was followed by John, then another John, who was one of the two bailiffs of the town in 1648, and then another William. In his time the property was divided, as it easily could be with the passage through the middle, though it remained in one ownership. The Hunts continued as ironmongers in one part; at one time a hatter traded in the other. In 1764 Thomas Hunt, ironmonger and brazier, is advertising in the Oxford Journal for a tenant for his adjoining shop. From the Hunts came one of Burford's most interesting men - the surgeon James Hunt, a pioneer of inoculation for smallpox in the 1770s, a great controversialist, a writer of medical texts and a formidable character who could and did on occasion see off an armed highwayman when he was travelling the lonely road between Burford and Northleach. James Hunt lived in a house on Church Green where the Warwick Hall is now but it was his grandson who, when the line of ironmongers came to an end, in 1823 sold the family house to Thomas Bowl, grocer. "Torpedo shaped" lemonade bottles of the sort that were closed with a marble can still be dug up in Burford gardens with the name Bowl moulded on them. Bowls the grocers continued to 1900 making a record of only two families and two trades in over three centuries. Then after the first war the Highway Hotel was opened by Miss Maddox in the building. In recent years several people have seen a young woman in Victorian dress with a frilled white apron quietly ascending the stair in this house. This peaceful apparition is typical of its centuries of settled domestic life and quiet prosperity. The hotel has closed, but now offers first-class "bed-and-breakfast" accommodation.

Above the Highway on the corner of Swan Lane is an old inn that has become a gallery. In 1652 it was the wealthy John Hanns' (or Hannys or Hannes) own house, but described as the **Greyhound** in 1680 (not to be confused with the Greyhound in Sheep Street). In 1685 it has become "*the Mansion house ... called the Greyhound and now in the several tenures of Hercules Hastings Robert West Richard Davis etc.*" (Hercules Hastings was a clock maker.) Knowing what we do of the good status and wealth of innkeepers and the earlier links of the Hanns family, it is likely that it was in business as the Greyhound in

1652 and earlier. Mr. John Hanns, Senior, was buried in 1660, and another John Hanns, surely his son, aged about 60, in 1667. Mrs. Hanns, "*widow, aged*" died in 1692, but it is probable that the house fell into multiple occupation before 1685. This Greyhound became the **Swan,** perhaps in the eighteenth century, though probably not overlapping in time with the Swan by the bridge, and remained so past the middle of this century. The name of Greyhound is later found in Sheep Street. The Swan gave its name to the lane leading from the High Street, which in the eighteenth century was known as Mullenders Lane, from an earlier family with land here.

Opposite was the curiously named **Three Goats' Heads**, in the property that is now the lower part of the Co-operative Shop. The first occurence of the name is in 1686, when Thomas Williams "*of the Goatsheads*" was buried. John Daniel was the keeper in the eighteenth century, and also ran a hemp dressing and sack weaving business. His son followed him, but his grandsons left Burford, one to be a Baptist missionary and the other to go to Luton to the straw hat trade. A grand-daughter, Abiah, remained and advertised for a manager for the business. James Wall came from Eynsham and added rope making to the business. Abiah married him, and with her straw hat connections she appears in the directories as milliner, as does her daughter after her. Somewhere along the line the inn disappeared, but J. Wall and Son, Rope Manufacturers, continued in business on the site. There is a passageway to the north of the property leading to the works erected behind in 1903 and in quieter days but still in living memory a boy twisting ropes might emerge and cross the High Street. Rope making ended after the war when the Co-operative Stores arrived, and other businesses use the works now.

Above is the **New Inn,** which is first found in the 1830s, though perhaps it was the **Quart Pot** earlier. The Quart Pot lasted through the eighteenth century and we know that it was in the High Street somewhere about here, though it may have moved about. John Eve, the keeper, was buried in 1696. Then from 1791 to 1802 it was in use for auctions as announced in the Oxford Journal. 1802 may have been the end: "*the Quart Pot to Lett - Mr. Roff is declining Public Business*". There are several business premises which look possibilities, with archways, yards or cellars but, more likely, the name was changed. The

New Inn was re-named the **Rampant Cat** after the last war. Robert FitzHamon, the Norman Lord of the Manor, who gave the town its charter around 1090, bore arms of a lion "rampant regardant". This was early adopted as the Corporation's seal, and later as the town's device, jestingly known as "the rampant cat". The Rampant Cat had a short day, for it is now the **Dragon,** a Chinese hostelry and restaurant.

Two seventeenth century inns are higher up the hill on this west side. The **Sun** is now Long Wivets, a whimsical name which is a traditional term for a size of stone roofing slate. The Sun first appears in 1685 and was in business through the eighteenth century. In 1793 it was for sale described as "Inn and Malt house", let on an annual tenancy. In 1819 it was occupied by Mr. Simpson, a grocer, so it may have ceased to be an inn, though the name lingered. It has been refronted this century.

The **Star** is a little older and is now Glenthorne, a private house. It is directly below the gap in the building line which once led to the "Thirty Acres" of the open fields of Burford, enclosed in the late seventeenth century, becoming the Star Close and later the Cow Pen Close, and now, owned by Burford Council since 1932, being the Recreation Ground and the West Field. The houses here on the hill are of seventeenth century date and the house may have been built as an inn. It is first mentioned in 1642 when John Taylor "*of the Star*" was buried. John Taylor had been bailiff of the Corporation in 1635, an important man in the town. We have his will, leaving his goods around the family and incidentally giving a picture of the inn. "*To my daughter, one bedstead in the Starre Chamber ... with all that belongeth to the same.*" We can picture a principal chamber in the inn with plasterwork of a star motif. The house itself went to his son Robert, who was a man of some property, and from him to another Robert, his son. This Robert, dying in 1695, may have been the last inn-keeper, for it ceased to be an inn around 1700. It came into the possession of Richard Whitehall, a substantial merchant. As an inn, the Star had presented two gables to the street, and probably under the northern one an archway led through the building, since cobbles have been found there. During the Whitehall ownership, it acquired its present frontage in the fashion of the time. Richard Whitehall himself lived on the other side of the hill in Simon Wisdom's house, which he also refaced.

A little higher up on the east side was the **White Horse**. From before 1636 this was the "capital messuage" of Thomas Silvester, clothier (i.e. cloth manufacturer) whose tomb is in the Lady Chapel of the Parish Church. His son, also Thomas, lived here for a while, but by 1685 it was divided into tenements, while the family, following the cloth trade, had moved to Witney. Around 1703 it passed out of Silvester ownership, and it was for sale in 1765, and before 1784 it was being used as a workhouse by the Upton and Signet overseers. It was a public house through the nineteenth century, but the large yard at the back was always an attraction for trades to be run in conjunction with the inn. Joseph Strafford, whose family was in the coaching business, ran it for thirty years across the turn of the century. In the 1840s William Green was a carpenter here, and in the 1860s Thomas Teago, who has the only cast iron grave "stone" in the churchyard, was in business as an agricultural machine maker. It was bought by Garne's Brewery in 1893 and, after the takeover by Wadworths, was closed as a public house recently. Houses have been built in the yard and the inn also is converted for private residence.

One more past inn remains on the hill, though perhaps ale-house would have been a better name. It is the **Chequers**, represented now by the house called Chippings, also at some time a row of cottages, and once containing a small meeting house. Many small sects have been represented in Burford over the centuries but, being quieter and more private, have left less of a mark in their passing than the ale-houses and inns.

We are now at the brow of the hill and from here, in body and in imagination, we can look back down the main road to the bridge far below us and in the mind's eye see the life of past centuries, the coaches and wagons jangling by and, as the light fades from the sky, the customers of the inns going about their business or taking their pleasure.

(Part Two of this survey will look at the inns of Burford beyond the High Street, and the story of the Burford Brewery.)

HindSight of Burford, 1996.

Index of Inns in Parts I and II of *The Inns of Burford*.

Andrews	H	*20*	I. 41
Angel	H	*16-18*	I. 10 II. 28
Angel	W	*20*	I. 10 II. 29
Bay Tree	S	*20*	II. 6-9,19
Bear	H	*15*	I. 10 II. 28
Bear	H	*17-20*	I. 18,30-32,35
Bell	H	*19*	I. 35,38-39
Bird in Hand	UR	*18-19*	I. 35 II. 36
Bird's Nest	UR	*19*	I. 35 II. 37
Blackamoor's Head		*17*	I. 41
Black Boy	W	*16-17*	II. 33
Black Horse		*18*	I. 35
Board and Letters		*18*	I. 35
Bull	H	*16-20*	I. 6,7,10,17,18-30,31,32,35,41
Burford House	H	*20*	I. 41
Burford Lodge	UR	*20*	II. 22
Catherine Wheel		*17*	I. 41
Chequers	H	*19*	I. 35,46
Coach and Horses	UR	*19*	I. 35 II. 37
Cooper's Addice(Adze)		*18*	I. 35
Corner House	H	*20*	I. 41
Cotswold Arms	H	*20*	I. 38
Cotswold Gateway	UR	*20*	II. 37
Cross Keys		*18*	I. 35
Crown	H	*(14)15-17*	I. 7-9
Dolphin	H	*18-19*	I. 35,39
Dragon	H	*20*	I. 45
Eight Bells	CL	*19*	I. 35 II. 34,35
(Golden) Fleece	S	*19*	I. 35 II. 25,26
George	H	*15-18*	I. 6,7,8,10-18,30,31,32,35 II. 18
Glaziers Arms		*18*	I. 35,36
Golden Ball	H	*19-20*	I. 35,40-41
Golden Ball	S	*18*	I. 40
Golden Pheasant	H	*20*	I. 40-41
Green Dragon		*18*	I. 41
Greyhound	H	*17*	I. 43-44
Greyhound	S	*19*	I. 35 II. 27
Half Moon		*17*	I. 41
High View	UR	*20*	II. 22,37
Highway	H	*20*	I. 42-43
Horse and Groom	W	*18*	II. 33
Horse and Jockey	W	*19*	I. 35 II. 33
Kings Arms	H	*18-19*	I. 35,36-37
Kings Head	W	*17-18*	I. 37 II. 34
Kings Head	H	*19*	I. 35,37
Lamb	S	*18-20*	I. 35 II. 9-17,19,23,24
Lamb and Flag	S	*19*	II. 25,26
Lamb (and Flag)	U+S	*18*	I. 35 II. 25,26

Masons Arms	W	*19-20*	I. 35 II. 29
Mermaid	H	*18-20*	I. 35,37-38
New Inn	H	*20*	I. 44
Old Mermaid	H	*18*	I. 35,37
Old Wheatsheaf	H	*18*	I. 35,42
Oxford	UR	*19*	II. 37
Plasterers Arms		*18*	I. 36
Plough	H	*19*	I. 39
Pound of Candles		*18*	I. 35
Quart Pott	H	*17-19*	I. 35,44
Rampant Cat	H	*20*	I. 45
Ramping Cat	UR	*19*	II. 37
Red Horse		*18*	I. 35
Red Lion	H	*18-19*	I. 35,40
Roebuck	W	*19*	II. 34
Rose and Crown	H	*18-19*	I. 35,39
Rose and Crown	U+S	*18-19*	I. 35 II. 36
Royal Oak	W	*18-20*	I. 35 II. 20,24,30
Star	H	*17-18*	I. 45
Sun	H	*17-19*	I. 35,45
Swan	H	*17-18*	I. 33-34
Swan	H	*19-20*	I. 35,44
Talbot	W	*17-18*	I. 20-21 II. 28
Three Cups		*17*	I. 41
Three Goats Heads	H	*17-18*	I. 44
Three Oranges		*18*	I. 35
Three Pigeons	H	*19-20*	I. 35,38
Three Sugar Loaves		*17*	I. 41
Three Towers	S	*18*	I. 35 II. 27
Unicorn	H	*18*	I. 35,36
Waggon & Horses		*18*	I. 35
Wheatsheaf	H	*18-19*	I. 35,42
White Hart	W	*17*	II. 30,31
White Hart	W	*18-19*	I. 35 II. 32
White Horse	H	*19-20*	I. 35,46 II. 24
Winter's Tale	UR	*20*	II. 37
Carpenters' Arms	F	*18-20*	II. 38
Dolphin	F	*18-20*	II. 38
Masons' Arms	F	*19*	II. 38
Red, White and Blue	F	*19*	II. 38
Tadpole	F	*18*	II. 38
Hit or Miss	T	*18*	II. 39
New Inn	T	*18*	II. 39
Bull, etc.	Stonelands	*18-19*	II. 39,40

H High Street; S Sheep Street; W Witney Street; CL Church Lane; UR Upper Road; U+S Upton and Signet; F Fulbrook; T Taynton. Italic numbers are centuries of known operation.